GCSE Physics Revise Guide

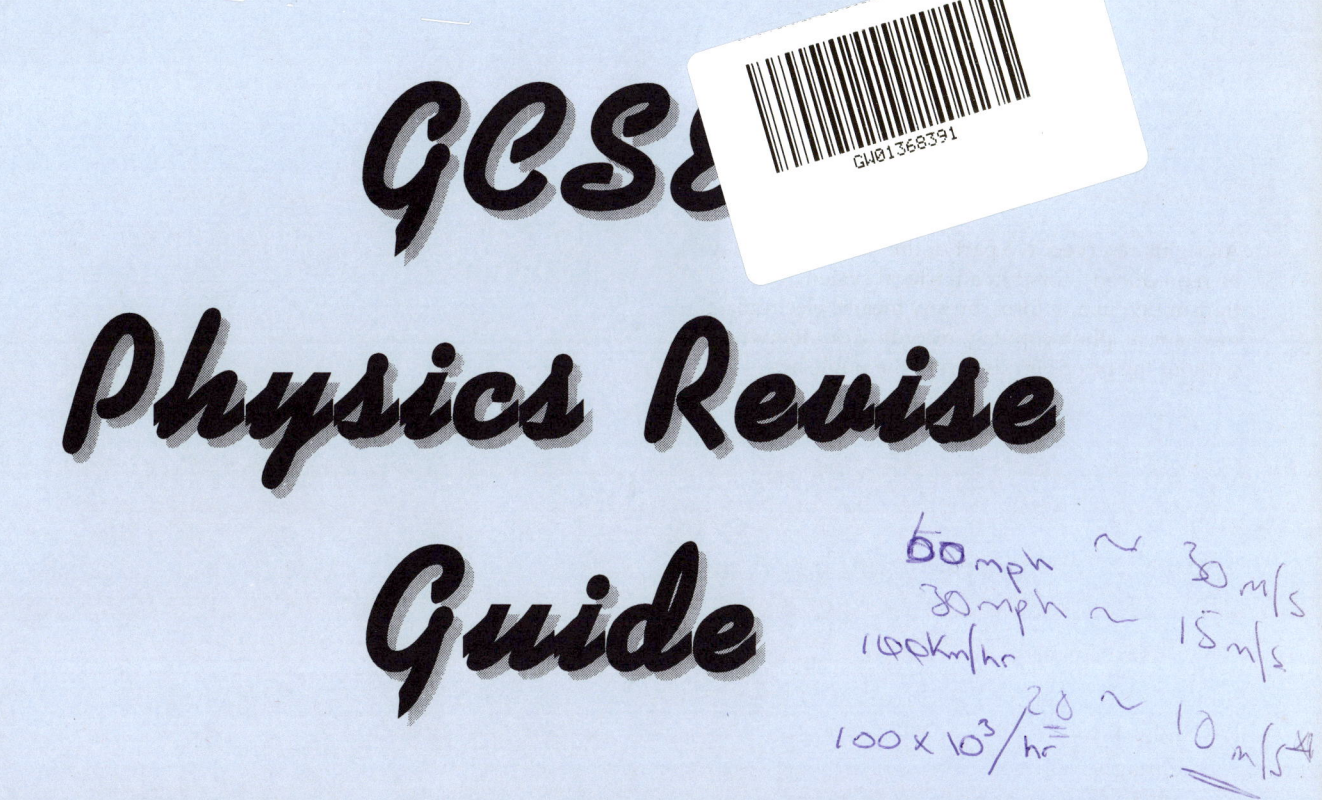

by

JOHN FRIEL

GCSE COURSE SUMMARY

FOR THE *NORTHERN IRELAND* CCEA SYLLABI:
"SCIENCE : PHYSICS"

AND

PHYSICS THEMES IN
"SCIENCE : DOUBLE AWARD"
(MODULAR & NON-MODULAR)

FOR 1997 EXAMS

All rights reserved. No part of this publication may be reproduced, stored in a retrieval system, or transmitted in any form, by any means, electronic, mechanical, photocopying, recording or otherwise, without the prior permission of the publisher.

© John Friel
  Omagh
  1997

ISBN 1 898392 36 6

COLOURPOINT BOOKS
Omagh Business Complex
Great Northern Road
OMAGH
Co TYrone
BT78 5LU

Tel: (01662) 250370
Fax: (01662) 249451

Introduction

As GCSE exam time approaches, many pupils (or their parents) buy revision guides. These guides cover the syllabi of all UK Examining Boards and they leave pupils with the difficult task of deciding which parts apply to the Northern Ireland CCEA syllabus.

This Revision Guide overcomes the problem in that it is written for the Northern Ireland syllabus only.

Due to curriculum changes, the summary applies to 1997 exams only, and it is my intention to produce a version which will apply from 1998 until the next syllabus revision (we are promised no changes for five years).

I would like to express my thanks to Ken Orr for his help in suggesting improvements to this publication.

I would welcome comments on this edition. These can be made either to Colourpoint or myself, care of the Colourpoint address.

John Friel,
December 1996

Important Note

Throughout the book, certain points are marked with this symbol. This indicates that the topic is NOT required knowledge for those studying Double Award Science.

Contents

A word about Units	5
Forces and Energy Formulae	6
Forces	7
Energy	10
Waves	13
Light	17
Electricity Formulae	22
Electricity and Magnetism	23
Radioactivity	27
The Earth in Space	29

Before you start ...
a word about Units

Before starting a calculation in Physics, it is a good idea to make a list of the numerical values given in the question.

If necessary before proceeding, values should be converted to standard units (metres, seconds, hertz, etc) without any prefixes such as *milli-*, *kilo-* or *mega-*:

eg. 1280 mm = 1.28 m, 2 kHz = 2,000 Hz

The only standard unit which always includes a prefix is kilogram (kg) for mass.

The following prefixes should be memorised:

M	mega....	10^6	× 1 000 000
k	kilo....	10^3	× 1 000
c	centi...	10^{-2}	× 0.01
m	milli...	10^{-3}	× 0.001
μ	micro...	10^{-6}	× 0.000 001
n	nano....	10^{-9}	× 0.000 000 001
p	pico...	10^{-12}	× 0.000 000 000 001

handwritten at top: Mol → 6.022 × 10²³ particles etc. / Amp → 6.24 × 10¹⁸ electrons

* not required for Double Award

FORCES & ENERGY FORMULAE

QUANTITY	UNITS	FORMULA
mass, m	kilograms, kg	
length, l	metres, m	
breadth, b	metres, m	
height, h	metres, m	
area, A	metres², m²	A = l x b
volume, V	metres³, m³	V = l x b x h
displacement, s	metres, m	
time, t	seconds, s	
velocity, v	m/s	v = s ÷ t
acceleration, a	m/s²	a = v increase ÷ t
retardation, a	m/s²	a = v decrease ÷ t
acceleration due to gravity, g	m/s²	
force, F	newtons, N	F = m a
weight, w	N	w = m g
momentum	kg m/s	momentum = m v
*density, ρ	kg/m³	ρ = m ÷ V
pressure, p	pascals, Pa	p = F ÷ A
moment	N m	force x perpendicular distance from pivot
work, W	joules, J	W = F x s *(ma×t)*
energy, E	joules, J	
kinetic energy, KE	joules, J	KE = ½ m v²
gravitational potential energy, GPE	joules, J	GPE = m g h *(g=10)*
Power, P	watts, W	P = W ÷ t P = E ÷ t
efficiency	%	eff = (W_out ÷ W_in) x 100 eff = (E_out ÷ E_in) x 100 eff = (P_out ÷ P_in) x 100

Forces

All FORCES are measured in Newtons (N).

An object is in TENSION when a force tends to increase its length in the direction of the force.

An object is in COMPRESSION when a force tends to decrease its length in the direction of the force.

The force of FRICTION tends to oppose motion and it can be an advantage (eg car brakes) or a disadvantage (eg slipping on ice).

Friction can be REDUCED by:

1) polishing surfaces

2) lubrication

3) streamlining (eg. cars, aeroplanes etc).

> **ELASTIC DEFORMATION** occurs in a material if it returns to its original shape after a stretching force has been removed.
>
> **PLASTIC DEFORMATION** occurs in a material if it retains its new shape after a stretching force has been removed.
>
> If a material is stretched beyond its **ELASTIC LIMIT**, then plastic deformation occurs.
>
> **HOOKE'S LAW** states that the extension of a material is directly proportional to the stretching force (or load), provided the elastic limit has not been exceeded.

The WEIGHT (N) of an object is the gravitational force exerted on it by the Earth or by another massive body.

MASS (kg) is an unchanging property of an object, whereas weight (N) depends on how strong gravity is.

A LEVER can act as a force multiplier (eg a crowbar).

The PRINCIPLE OF MOMENTS states that when an object is in equilibrium, the anticlockwise moment about any point is equal to the clockwise moment about that point.

If an object, which is acted on by several forces, is in EQUILIBRIUM:

(1) the sum of the forces in one direction must equal the sum of the forces in the opposite direction,

(2) the principle of moments must apply.

The CENTRE OF MASS (or centre of gravity) of an object is the point at which all of the object's mass (or weight) appears to be concentrated.

An object will tend to TOPPLE if it is moved to a position where its centre of mass is no longer above its base.

An object (such as a racing car) will have good STABILITY if it has:

　　1) a low centre of mass,　　2) a wide base.

A circular cone is in:

　　1) STABLE EQUILIBRIUM when it rests on its base,

　　2) UNSTABLE EQUILIBRIUM when it is balanced on its point,

　　3) NEUTRAL EQUILIBRIUM when it lies on its side.

DENSITY is a characteristic property of a material, which means that the density always stays the same, regardless of the size of the piece of the material.

"A DENSITY OF 7 g/cm³" means that every 7 g mass of the material has a volume of 1 cm³.

An object will FLOAT in a liquid if its average density is less than that of the liquid, and it will sink in a liquid if its average density is greater than that of the liquid.

(The DENSITY OF WATER is 1,000 kg/m³ or 1 g/cm³).

One PASCAL is the pressure exerted by a force of 1 N on an area of 1 m² (1 Pa = 1 N/m²). For a set force, a small area gives a large pressure and a large area gives a small pressure.

When an effort is applied to the small cylinder in a HYDRAULIC JACK, the fluid becomes pressurised, and the pressure is transmitted to the piston in the large cylinder, enabling it to raise a load which is greater than the effort.

AVERAGE SPEED (m/s) = total distance (m) ÷ total time (s)

AVERAGE SPEED (m/s) = {initial speed + final speed} ÷ 2.

RATE OF CHANGE OF SPEED (m/s²) = {final speed - initial speed} ÷ time.

DISPLACEMENT is distance (m) in a certain direction.

VELOCITY is speed (m/s) in a certain direction and it equals rate of change of displacement.

[NB: *In Double Award Science, distance and displacement are taken to mean the same thing, as are speed and velocity.*]

ACCELERATION equals rate of change of velocity.

ACCELERATION (m/s²) = {final velocity - initial velocity} ÷ time.

"AN ACCELERATION OF 6 m/s²" means that an object's velocity increases by 6 m/s in every second.

DISPLACEMENT v TIME GRAPH,　　gradient = velocity.

VELOCITY v TIME GRAPH,　　gradient = acceleration,

　　　　　　　　　　　　　　　negative gradient = deceleration or retardation,

　　　　　　　　　　　　　　　area under line = displacement.

EQUATIONS OF MOTION,

s = displacement (m), t = time (s), u = initial velocity (m/s)

v = final velocity (m/s), a = acceleration (m/s²).

1) $v = u + at$,   2) $s = \frac{1}{2}(v + u)t$,   3) $s = ut + \frac{1}{2}at^2$,   4) $v^2 = u^2 + 2as$.

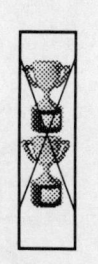

NEWTON'S FIRST LAW:	**NEWTON'S SECOND LAW:**	**NEWTON'S THIRD LAW:**
An object will continue in a state of rest or uniform motion in a straight line unless a resultant force acts on it.	The resultant force (N) acting on a body equals its mass (kg) times its acceleration (m/s²), or $F = ma$.	The force exerted by a body A upon a body B is equal in magnitude to, opposite in direction to, and in the same straight line as the force exerted by B on A, **or** to every action, there is an equal and opposite reaction.

When the forces acting on an object are not balanced, the net force is called the RESULTANT FORCE.

One NEWTON is the force which gives a mass of 1 kg an acceleration of 1 m/s².

In the absence of other forces, all bodies close to the earth fall with the same acceleration, which is called the ACCELERATION DUE TO GRAVITY, g, or the acceleration of 'free fall'.

CENTRIPETAL FORCE is the force needed to make an object move in a circle, and it is always directed towards the centre of the circle.

A LARGER centripetal force is needed if:

1) the mass of the object is increased,

2) the speed of the object is increased, or

3) the radius of the circle is decreased.

$$F_c = \frac{mv^2}{r}$$

If the centripetal force is removed, the object will fly off at a TANGENT to the circle.

A gun RECOILS when fired as the gun has an equal and opposite momentum to the bullet. The bullet has a small mass and a large velocity while the gun has a large mass and a small velocity.

ROCKET and JET engines produce exhaust gases with momentum backwards, which results in the engines having an equal momentum forwards.

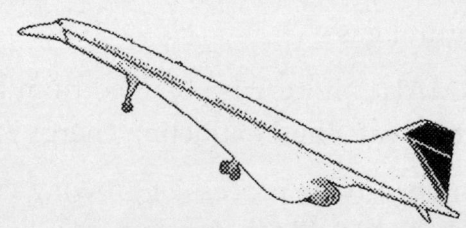

Energy

ENERGY, measured in Joules (J), is the amount of work that a body is capable of doing.

The main FORMS OF ENERGY are:

Gravitational Potential (due to position above the ground),

Strain Potential (due to condition, eg. a wound spring),

Kinetic (due to movement),

Chemical (as stored in food and fuels),

Heat, Light, Sound, Electrical, Magnetic, Nuclear.

A MACHINE is a device to make the transfer of energy more convenient, and it can be either a force multiplier (eg. a crowbar) or a distance multiplier (eg. the lower arm bone).

The LAW OF CONSERVATION OF ENERGY states that energy cannot be created or destroyed, but it can be changed from one form to another.

FRICTION is a mechanism by which kinetic energy is transformed into heat energy.

The ultimate result of ENERGY TRANSFERS is to increase the temperature of the surroundings and useful energy is dissipated as heat.

A RENEWABLE source of energy (eg.wind) will last as long as the Sun while a non-renewable source (eg.oil) will eventually be used up.

The SUN is the major energy source for the earth.

NUCLEAR FISSION occurs when the nucleus of an atom splits into two or more smaller nuclei, releasing energy. The fuel is usually uranium.

NUCLEAR FUSION (the reaction which takes place in the Sun) occurs when two hydrogen nuclei fuse together to form one helium nucleus, releasing energy.

A FOSSIL FUEL is one which is formed from dead, decayed plants and animals, eg coal, oil, peat, natural gas.

GEOTHERMAL ENERGY is obtained from the heat below the crust of the earth by pumping down cold water and extracting energy from the hot water which comes back up.

BIOMASS refers to the process of growing crops for energy, eg. sugar cane in Brazil from which alcohol is obtained.

Energy changes in generating electricity in a CONVENTIONAL POWER STATION:

fuel *(chemical)* ⟶ furnace *(heat)* ⟶ boiler *(strain potential)* ⟶ steam turbine *(kinetic)* ⟶ alternator *(electrical)*

Energy changes in generating electricity in a NUCLEAR POWER STATION:

uranium *(nuclear)* ⟶ reactor core *(heat)* ⟶ boiler *(strain potential)* ⟶
steam turbine *(kinetic)* ⟶ alternator *(electrical)*

EFFICIENCY (%) is a measure of how much energy is transferred in an intended way.
Efficiency = (Energy out ÷ Energy in) x 100.

The OUTPUT ENERGY of a machine is always less than the input energy, as a machine's efficiency is always less than 100%, because of heat energy lost due to friction.

HEAT is the form of energy which flows from a hot object to a cool object.

TEMPERATURE is a measure of the average kinetic energy which each molecule of an object possesses.

The LOWER FIXED POINT (0 degrees Celsius) is the melting point of pure ice.

The UPPER FIXED POINT (100 degrees Celsius) is the boiling point of pure water.

CONDUCTION occurs when heat flows to try to reduce a temperature difference in a material without movement of the material itself (eg. heat flowing through a metal poker which has one end in a fire).

A material which is a poor conductor of heat is called an INSULATOR.

U-VALUES refer to the insulating ability of a material. The lower the U-value, the better the insulator.

The U-VALUE of a material (measured in W/m^2 ^{0}C) is the rate at which heat loss (in J/s or W) occurs through an area of 1m^2 of the material when there is a temperature difference of 1^{0}C.

U-VALUE EQUATION: $E/t = U A (\theta_2 - \theta_1)$,

where: **E/t is rate of heat loss in J/s or W,**

U is the U-value in W/m^2 ^{0}C,

A is the area in m^2,

θ_2 is the inside temperature in ^{0}C,

θ_1 is the outside temperature in ^{0}C.

CONVECTION occurs when liquids and gases (which are poor conductors) carry heat from one place to another if they are free to circulate.

RADIATION occurs when infrared waves travel through a vacuum or through gases (without heating the gases up).

A black surface is a better ABSORBER and EMITTER of heat radiation than a shiny or white surface.

The GREENHOUSE EFFECT occurs when short wavelength infrared radiation from the Sun passes through glass and heats up materials inside. These then radiate long wavelength infrared radiation which reflect off the glass. Hence, the inside of the greenhouse becomes warmer.

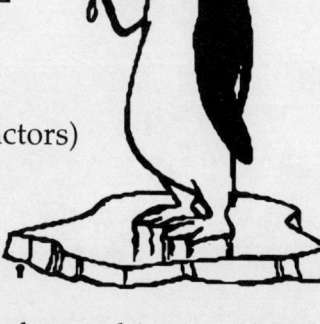

CARBON DIOXIDE GAS in the atmosphere acts like glass in a glasshouse and traps the heat of the Sun. Burning of fossil fuels is increasing the amount of CO_2 in the atmosphere, leading to global warming.

WORK is done when the point of application of a force moves along its line of action.

One JOULE of work is done when when the point of application of a force of 1 newton moves through a distance of 1 metre.

POWER is the rate at which work is done, or at which energy is converted from one form to another.

One WATT is the power exerted when one joule of work is done in one second.

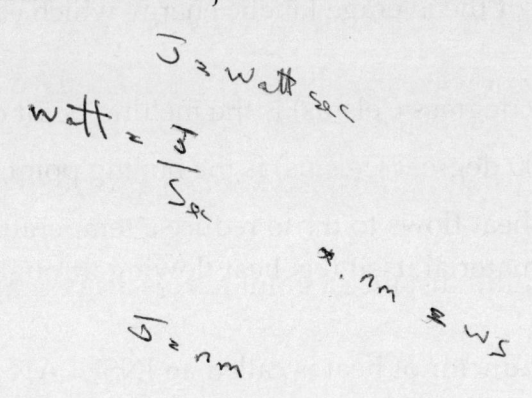

$$J = Watt \cdot sec$$
$$Watt = \frac{J}{sec}$$
$$W = \frac{J}{m}$$
$$J = nm$$

Waves

WAVES transfer energy as they travel from one place to another.

In a TRANSVERSE WAVE, the oscillations take place at right angles to the direction of travel (eg. water and electromagnetic waves).

In a LONGITUDINAL WAVE, the oscillations take place in the direction of travel (eg. sound).

PERIODIC TIME, T, (s) is the time for one complete wave to pass a point or for one complete oscillation to take place.

FREQUENCY, f, is the number of waves passing a point in one second, measured in waves per second or Hertz (Hz).   Frequency = 1 / Periodic Time. $F = \frac{1}{T}$

WAVELENGTH, λ, (m) is the distance from any point on a wave to the same point on the next wave.

AMPLITUDE (m) is the maximum distance a point moves away from its rest position when a wave passes.

The WAVE EQUATION: $v = f\lambda$, where v is the wave velocity (in m/s).

A vibrating object has a characteristic FUNDAMENTAL FREQUENCY of vibration.

RESONANCE (a vibration of large amplitude) occurs when an object is made to vibrate at its fundamental or natural frequency.

RESONANT OSCILLATIONS can be advantageous (eg. musical instruments) or disadvantageous (eg. Tacoma Narrows Bridge).

REFRACTION occurs when a wave crosses the boundary between different media.

When a water wave passes into shallower water, its speed and wavelength decrease while its frequency remains constant, and vice versa:

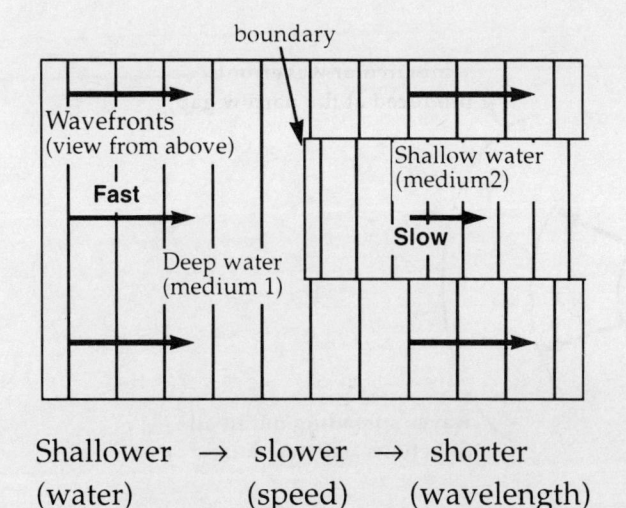

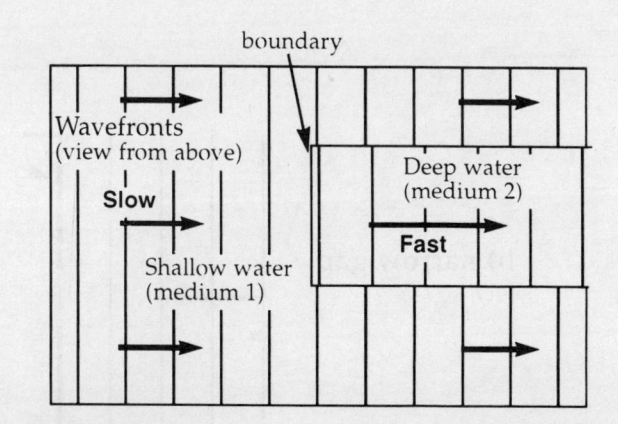

Shallower → slower → shorter
(water)   (speed)   (wavelength)

If the DIRECTION of the wave is not at right angles to the boundary, the direction changes as shown:

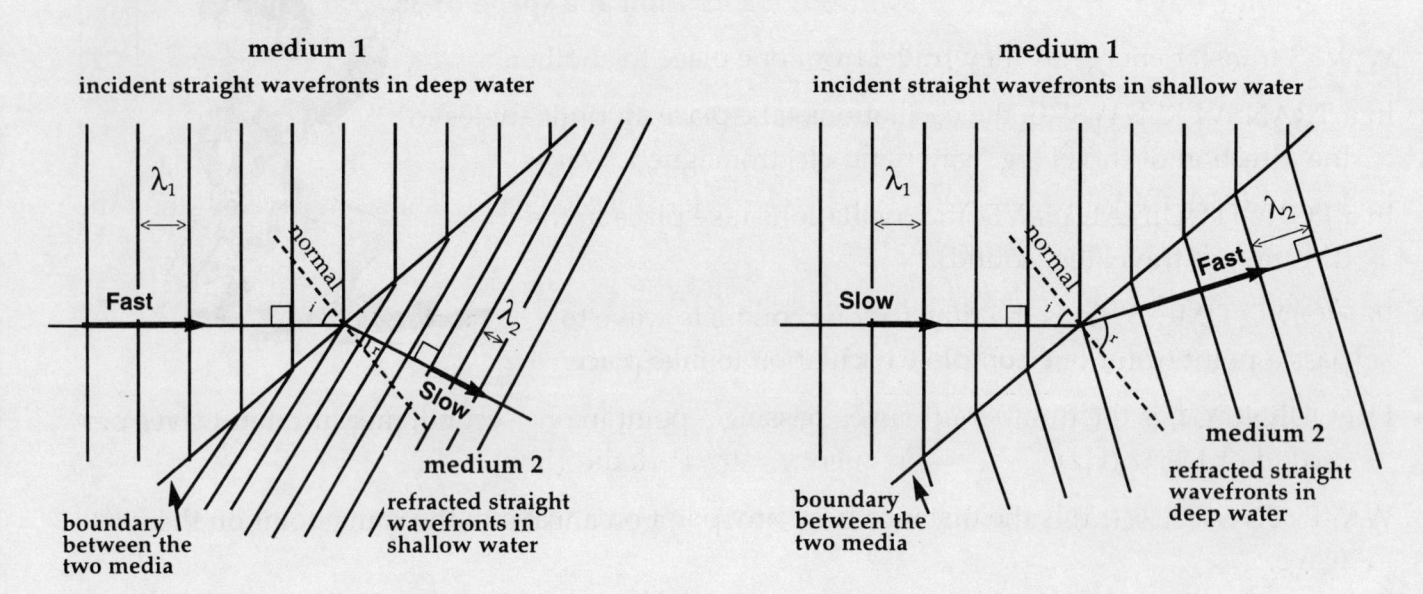

medium 1 — incident straight wavefronts in deep water

medium 1 — incident straight wavefronts in shallow water

DIFFRACTION is the bending of waves as they pass around obstacles. When plane waves pass through a gap which is comparable with wavelength of the waves, a semi-circular wave pattern is produced:

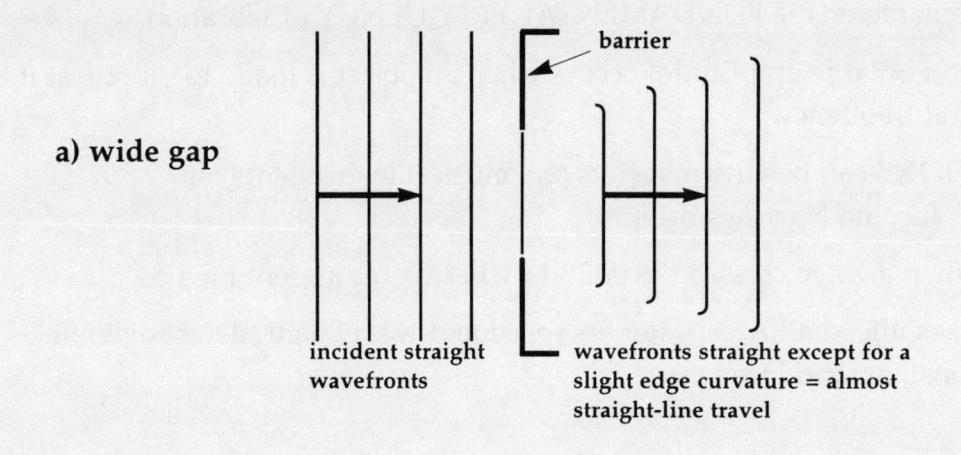

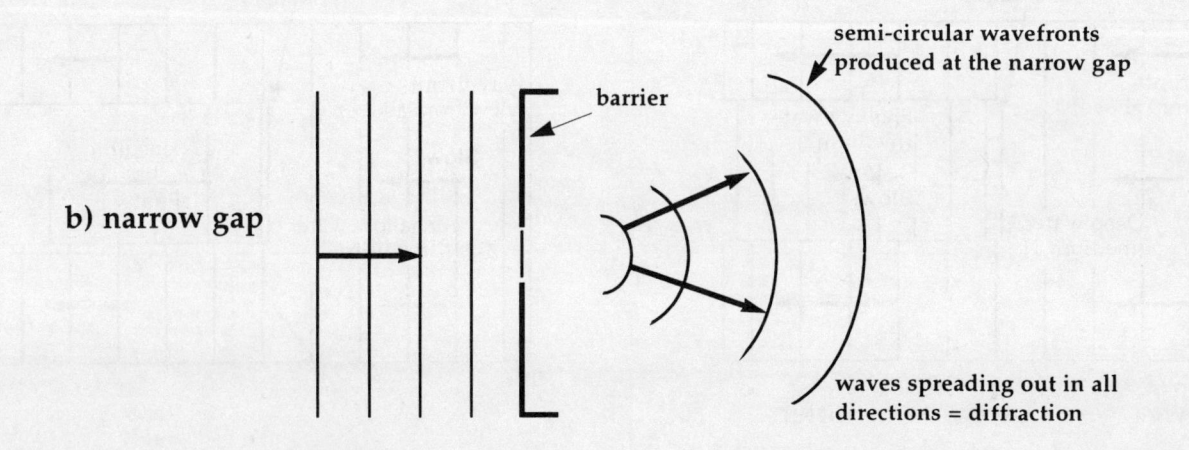

a) wide gap

incident straight wavefronts — wavefronts straight except for a slight edge curvature = almost straight-line travel

b) narrow gap

semi-circular wavefronts produced at the narrow gap

waves spreading out in all directions = diffraction

A POLARISED WAVE is one which vibrates in one plane only.

ELECTROMAGNETIC WAVES have the following properties:
1) they require no medium,
2) they travel in straight lines through a vacuum at a speed of 300,000 km/s (known as the speed of light),
3) they are transverse.

In the ELECTROMAGNETIC SPECTRUM, the order from longest to shortest wavelength (and from lowest to highest frequency) is:

Radio   Microwave   Infra-red   Light   Ultra-Violet   X Rays   Gamma Rays

SOUND is a longitudinal wave produced by a vibrating source.

Sound requires a MEDIUM, either solid, liquid or gas, to travel through and it cannot travel through a vacuum.

The SPEED OF SOUND in liquid is greater than in a gas, and in a solid it is greater than in a liquid.

REFRACTION occurs when a sound wave crosses the boundary between media of different densities. When it travels into a less dense medium, its speed and wavelength decrease while its frequency remains constant, and vice versa

Sound passing into a MORE DENSE medium is bent away from the normal (angle of incidence is less than the angle of refraction):

Sound passing into a LESS DENSE medium is bent towards the normal (angle of incidence is greater the angle of refraction):

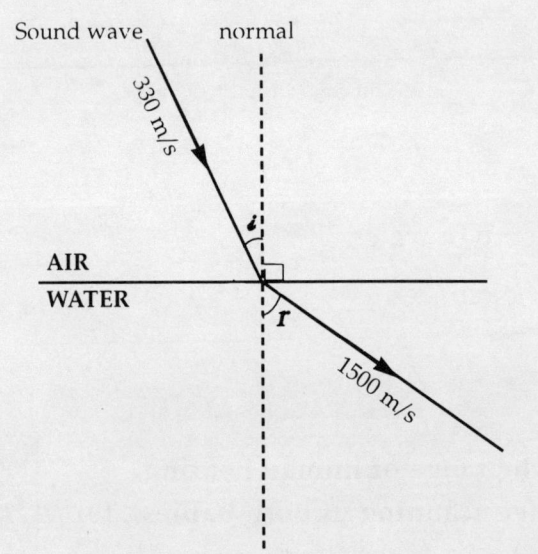

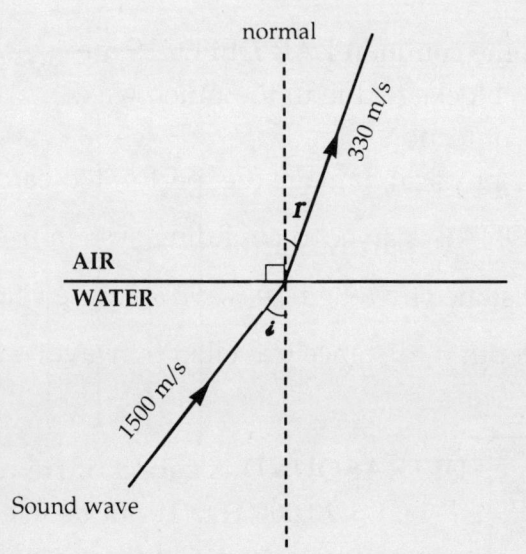

i angle of incidence
r angle of refraction

The LOUDNESS of a sound increases if the amplitude of the vibration increases.

The PITCH of a sound increases if the frequency of the vibration increases (and the wavelength decreases).

The EAR has three regions:

1) OUTER EAR: sounds enter the ear and make the eardrum vibrate,

2) MIDDLE EAR: sounds pass through the three bones (hammer, anvil and stirrup) which amplify the sounds 22 times.

3) INNER EAR: sounds pass via the oval window into the fluid filled spiral canal called the cochlea. This converts sounds into electrical signals which travel through nerves to the brain.

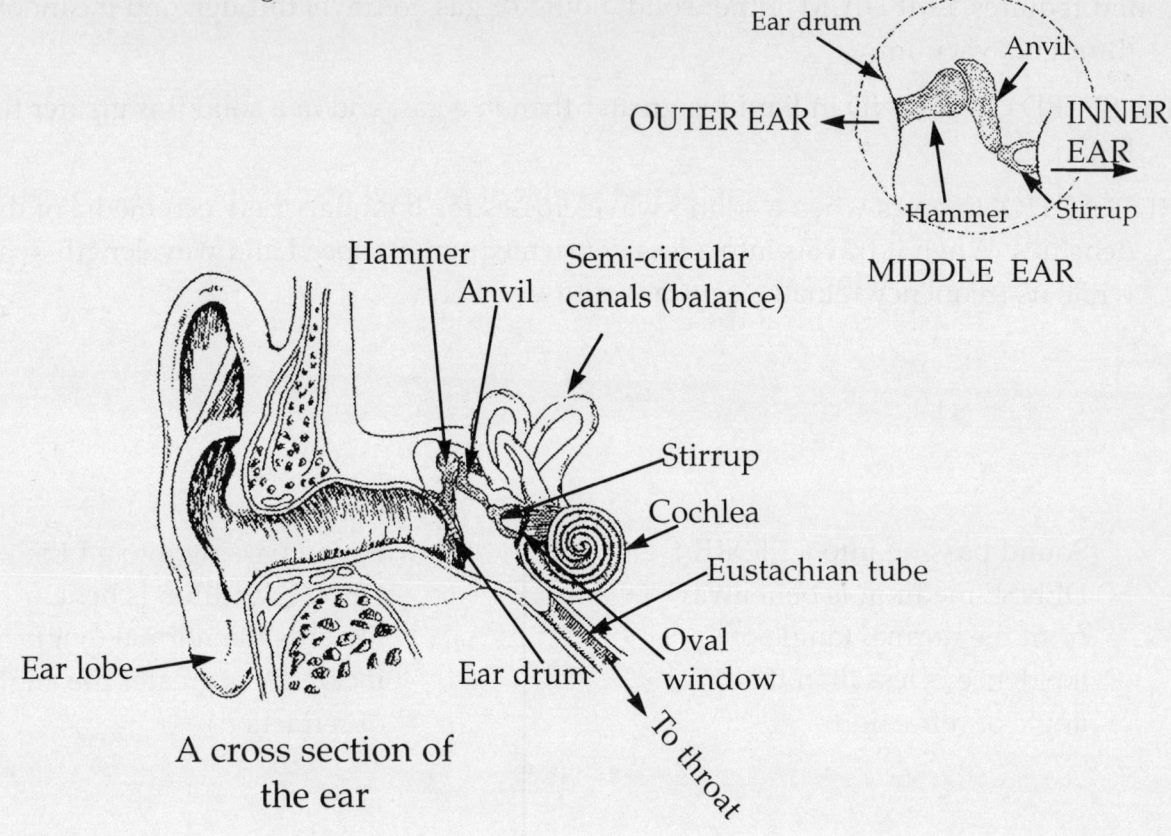

A cross section of the ear

Some common EAR DEFECTS are:
blockage due to too much wax,
infection
physical damage (eg. perforated eardrum).

SONAR is an echo-sounding system used to find the depth of water.

Distance travelled by wave = (wave velocity) X (time taken).

Depth = (distance travelled by wave) ÷ 2

ULTRASOUND is sound of frequency beyond the range of human hearing, ie. f > 20,000 Hz. It can be used in medicine (eg scanning unborn babies), for crack detection and for cleaning objects.

Light

LIGHT is the form of energy to which the retina of the eye is sensitive.

An UMBRA is a shadow with sharp edges caused by a point source of light and PENUMBRA is the name of the area of part shadow surrounding the umbra caused by an extended source of light.

A REAL IMAGE is one which can be formed on a screen, while a VIRTUAL IMAGE (or IMAGINARY IMAGE) is one which cannot be formed on a screen.

A NORMAL is a line drawn at right angles to a mirror and angles of incidence, reflection and refraction are measured between the rays and the normal.

The LAW OF REFLECTION states that when a ray of light strikes a reflecting surface, the angle of incidence equals the angle of reflection.

IMAGE RULES: When an object is placed in front of a plane (or flat) mirror, the image formed is:
   1) **the same size as the object**
   2) **as far behind the mirror as the object is in front**
   3) **virtual**
   4) **laterally inverted**

The SPEED OF LIGHT in a gas is greater than in a liquid, and in a liquid it is greater than in a solid.

REFRACTION occurs when a light ray crosses the boundary between media of different densities. When it travels into a more dense medium, its speed and wavelength decrease while its frequency remains constant, and vice versa.

Light passing into a MORE DENSE medium is bent towards the normal (angle of incidence is greater than the angle of refraction).	Light passing into a LESS DENSE medium is bent away from the normal (angle of incidence is less than the angle of refraction):

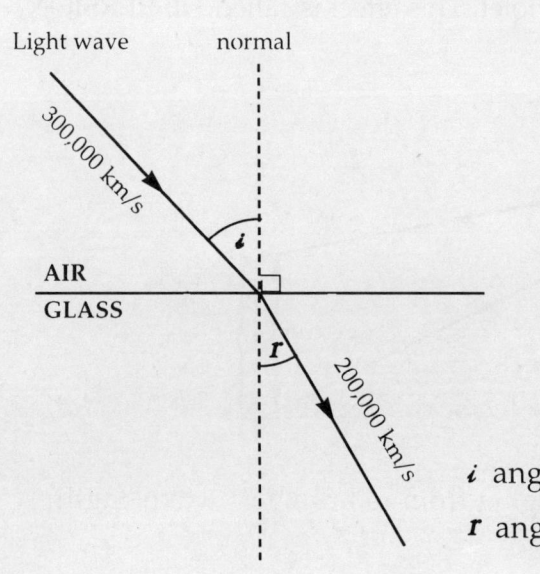

i angle of incidence

r angle of refraction

The REAL DEPTH of a pond is the actual depth, while the APPARENT DEPTH is the depth of the image of the bottom of the pond, due to refraction of light rays at the surface of the water:

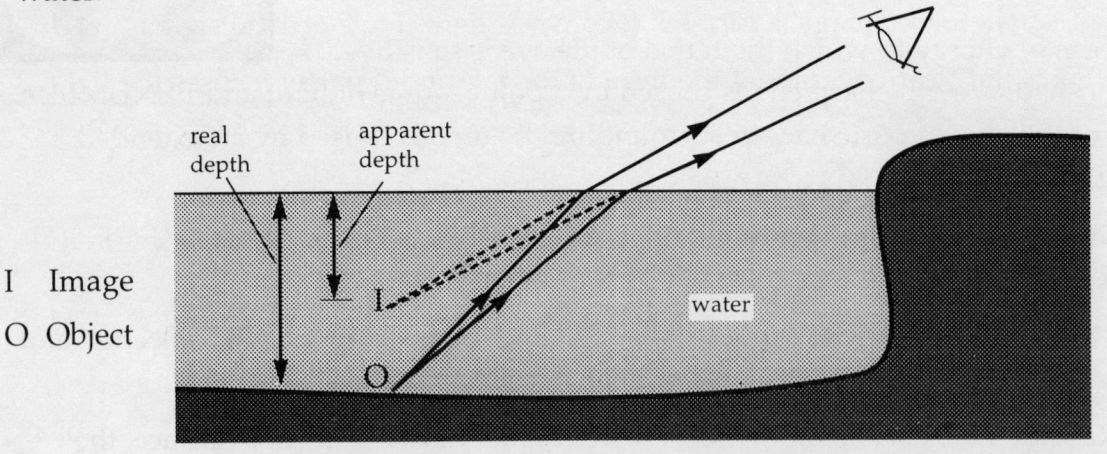

I Image
O Object

TOTAL INTERNAL REFLECTION occurs when light travels to a water, glass or perspex/air boundary at an angle of incidence greater than the critical angle for the material. All of the ray is reflected and none is refracted.

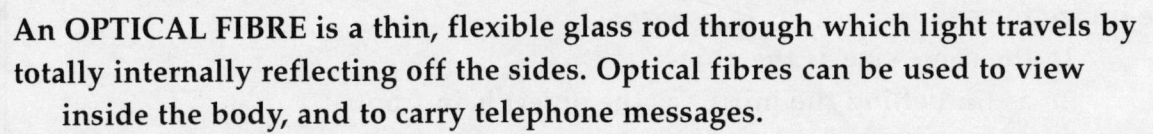

An OPTICAL FIBRE is a thin, flexible glass rod through which light travels by totally internally reflecting off the sides. Optical fibres can be used to view inside the body, and to carry telephone messages.

Normal light vibrates at right angles to the direction of travel in many planes, but POLARISED light vibrates in just one plane

Reflected light is mainly horizontally polarised, while SUNGLASSES only allow vertically polarised light through. Thus, sunglasses reduce glare.

If a parallel beam of light passes through a CONVEX LENS, the beam *converges* towards a point called the PRINCIPAL FOCUS.

If a parallel beam of light passes through a CONCAVE LENS, the beam *diverges* as if it has come from a point called the PRINCIPAL FOCUS.

The FOCAL LENGTH of a lens is the distance from the principal focus to the centre of the lens.

In a SPECTRUM produced by shining white light through a prism, the least deviated colour is red, followed by orange, yellow, green, blue, indigo and violet. This effect is called DISPERSION.

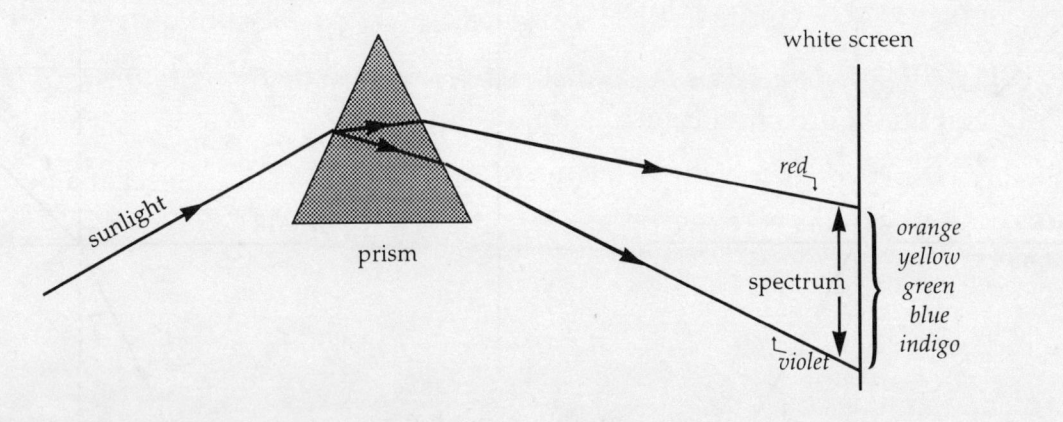

The COLOUR of light depends on its wavelength, ranging from red (longest wavelength) through to violet (shortest wavelength).

The functions of the parts of the EYE

- The SCLEROTIC is the skin which surrounds the eye.
- The RETINA is where the image is formed. It converts light into electrical signals.
- The CORNEA is the clear covering at the front of the eye and it partly focuses the light entering.
- The PUPIL is the hole through which light enters the eye.
- The IRIS changes the size of the pupil making it smaller for bright light or larger for dull light.
- The LENS focuses the image on the retina.
- The CILIARY MUSCLES adjust the shape of the lens for focusing.
- The OPTIC NERVE carries the electrical signals from the retina to the brain.
- The BLIND SPOT is the point where the optic nerve is connected to the retina.

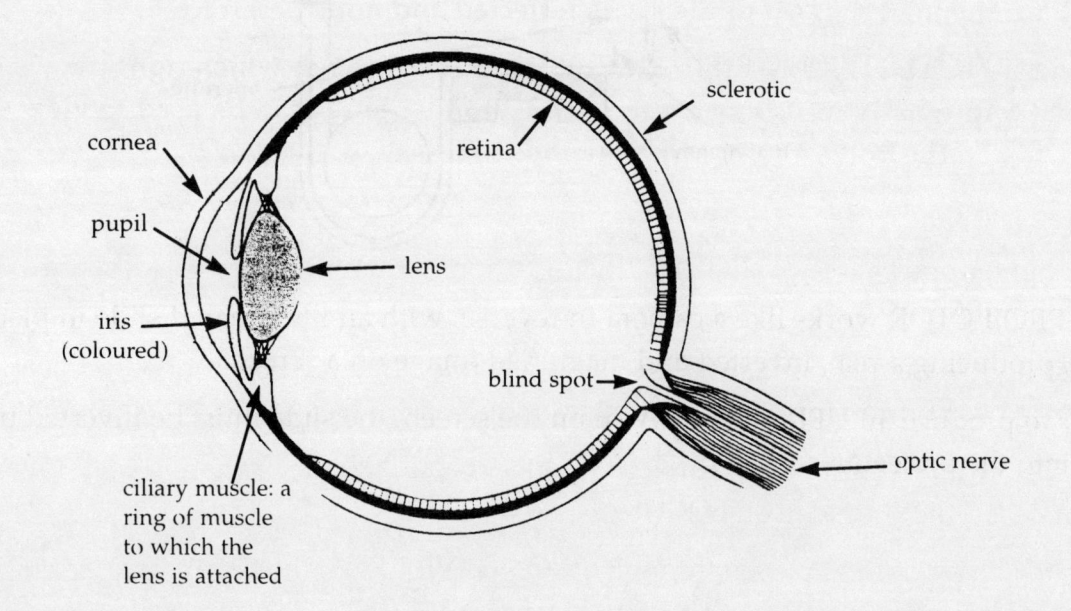

ACCOMMODATION is the ability of the eye lens to change shape in order to focus on near or far objects. For near objects, the lens is fat (small focal length) while for far objects, the lens is thin (large focal length).

The nature of the IMAGE ON THE RETINA is: real, inverted and diminished.

SHORT SIGHT: the eye lens focuses parallel rays from a distant object to form an image in front of the retina. Correction is by a *concave* lens.

LONG SIGHT: the eye lens tries to focus diverging rays from a close object at a point behind the retina. Correction is by a *convex* lens.

The functions of the parts of the CAMERA

- The CONVEX LENS is moved towards the film to focus the image of far objects on the film, and is moved away from the film to focus near objects.
- The APERTURE is the hole through which light passes into the camera.
- The DIAPHRAGM changes the size of the aperture smaller for bright light or larger for dull light.
- The SHUTTER is opened for a very short time to allow light in while a photograph is being taken.
- The EXPOSURE TIME is the length of time for which the shutter is open.

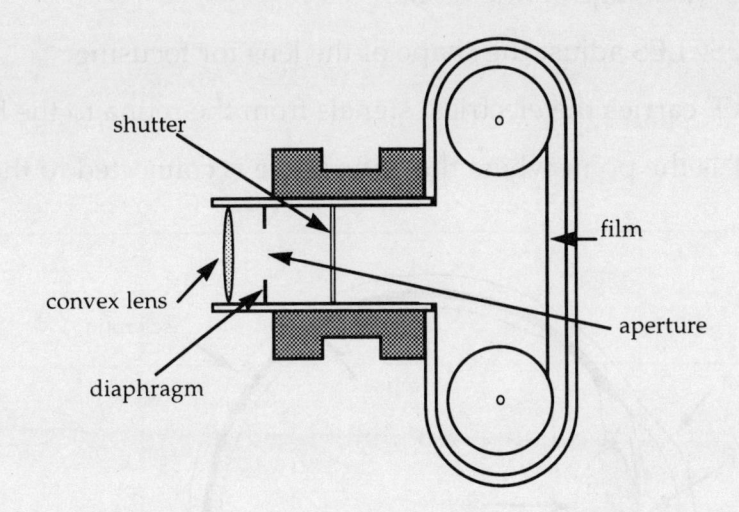

A slide PROJECTOR works like a camera in reverse, with an illuminated slide in place of the film, producing a real, inverted and magnified image on a screen.

In order to produce an UPRIGHT IMAGE on the screen, the slide must be inverted before it is put into the projector.

CONVEX LENS RAY DIAGRAMS

This diagram illustrates number 4 in the chart below. **O** = Object; **I** = Image

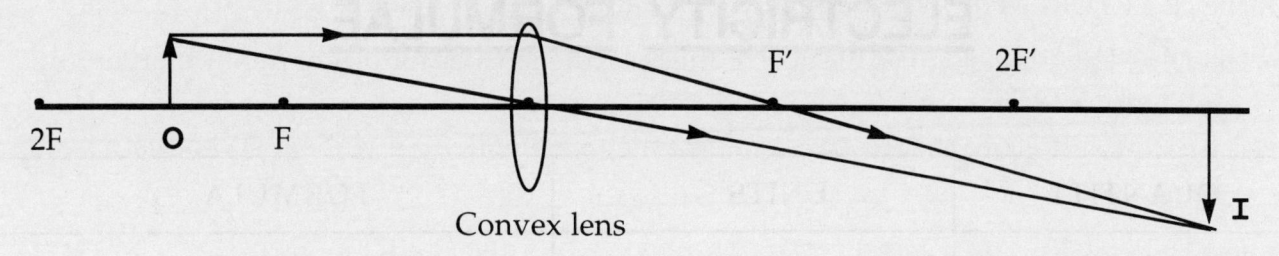

Convex lens

F and F' are principal foci and the line through them is the principal axis.

	OBJECT	IMAGE	USE
1	* At infinity	Real, inverted, diminished, at F'	Approximate measurement of focal length
2	Between infinity & 2F	Real, inverted, diminished, between 2F' & F'	Camera and eye
3	At 2F	Real, inverted, same size as object, at 2F'	Camera making equal size copies
4	Between 2F & F	Real, inverted, magnified, between 2F' & infinity	Projector and enlarger
5	* At F	At infinity	Accurate measurement of focal length
6	* Between F & lens	Virtual, upright magnified, between object & infinity	Magnifying glass

* not required for Double Award

ELECTRICITY FORMULAE

QUANTITY	UNITS	FORMULA
length, l	metres, m	
radius, r	metres, m	
area, A	metres^2, m^2	$A = \pi r^2$
time, t	seconds, s	
current, I	amperes, A	
charge, Q	coulombs, C	$Q = It$
potential difference, V	volts, V	
resistance, R	ohms, Ω	$R = V/I$
* resistivity, ρ	Ωm	$\rho = RA/l$
resistors in series	Ω	$R = R_1 + R_2$
resistors in parallel	Ω	$1/R = 1/R_1 + 1/R_2$
power, P	watts, W	$P = VI = I^2R = V^2/R$
energy, E	joules, J	$E = Pt = VIt = I^2Rt = V^2t/R$
energy, E	kilowatt-hours, kWh	$E = Pt$ (P in kW, t in hours)

* not required for Double Award

Electricity and Magnetism

Inside the atom, ELECTRONS have a negative (-) charge, PROTONS have an equal positive (+) charge and NEUTRONS have no charge.

LIKE charges repel and UNLIKE charges attract.

ELECTRICAL CURRENT consists of the movement of free electrons.

CONVENTIONAL current direction, from positive to negative, is opposite to the direction of electron flow.

A COMPLETE CIRCUIT is needed for an electrical device (eg. a bulb) to work, ie. the electricity must be able to get back to where it started from.

DIRECT CURRENT does not change in size or direction whereas ALTERNATING CURRENT continually changes in both size and direction in a regular pattern.

The frequency of MAINS ELECTRICITY SUPPLY is 50 hertz.

An ELECTROSTATIC CHARGE builds up in an object when friction causes it to gain extra electrons (making it negatively charged) or to lose electrons (making it positively charged), eg. a person walking on a certain type of carpet.

A LIGHTNING CONDUCTOR allows lightning (electricity originating in the clouds) to flow safely down the side of a tall building and into the earth.

CONDUCTORS (all metals and graphite) are materials which allow electrons to flow through them because their own electrons are FREE to move around.

INSULATORS (eg. wood, plastic, glass, rubber, paper) are materials which do not allow electrons to flow through them because their own electrons are BOUND to their atoms.

POTENTIAL DIFFERENCE is measured in volts and electricity always flows from a point at higher potential to a point at lower potential.

A VOLTMETER is used to measure the potential difference across a component and it is placed in parallel with the component.

PRIMARY CELLS irreversibly convert chemical energy to electrical energy.

SECONDARY CELLS reversibly convert electrical energy to chemical energy, for storage purposes.

One COULOMB of charge consists of approximately 6 million million million electrons.

One COULOMB of charge passes a point in one second when a current of one AMPERE is flowing.

An AMMETER is used to measure the current flowing through a component and it is placed in series with the component.

The current is the same at all points in a SERIES circuit.

The sum of the potential differences in a SERIES circuit is equal to the potential difference across the whole circuit.

The sum of the currents in the branches of a PARALLEL circuit is equal to the current entering the parallel section.

OHM'S LAW

The current (A) flowing through a metal conductor is directly proportional to the potential difference (V) across its ends, provided its temperature remains constant.

A conductor has a resistance of one OHM (Ω) if a current of one amp flows through it when a potential difference of one volt is applied across its ends.

The gradient of a POTENTIAL DIFFERENCE - CURRENT GRAPH for a metallic conductor at constant temperature equals the conductor's resistance in Ω.

The RESISTANCE OF A METALLIC CONDUCTOR is directly proportional to the length of the conductor and inversely proportional to its cross sectional area.

The RESISTIVITY of a material (measured in Ωm) is equal to the resistance of 1 m of the material of cross sectional area 1 m^2.

THREE PIN PLUG

right hand terminal, live wire - brown

left hand terminal, neutral wire - blue

centre terminal, earth wire - green/yellow

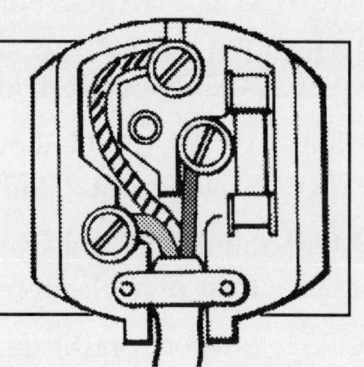

The ELECTRICAL POWER of a device (measured in watts) is the rate at which electrical energy is converted into other forms of energy.

One KILOWATT HOUR (kWh or UNIT) is the energy supplied in 1 hour to an appliance whose power is 1 kW. It is equivalent to 3.6 MJ

In a RADIAL CIRCUIT, 13 amp sockets are connected in series to a consumer unit (or fusebox).

In a RING CIRCUIT, 13 amp sockets are connected to the consumer unit by a loop so that there are two conducting paths to each socket.

A FUSE contains a short piece of thin wire which overheats and melts when too high a current flows through it. It is placed in series with the live wire.

The SIZE OF FUSE needed for a device is found by calculating the current using : I = P/V. The fuse is then selected from: 3A, 5A or 13A.

A CIRCUIT BREAKER (sometimes used instead of a fuse) is an automatic switch which opens when too high a current flows through it.

An electrical device has DOUBLE INSULATION if it has a plastic body, and hence does not need an earth lead.

FUSE RATINGS for different household circuits are:
- cooker — 30 A,
- ring main circuit — 30 A,
- radial main circuit — 20 A,
- immersion heater — 15 A,
- lighting circuit — 5 A.

The MAGNETIC FIELD pattern, due to a current flowing through a STRAIGHT WIRE, is a series of concentric circles. The field direction is the same as that in which a right handed screwdriver is turned to move a screw the same way as the current.

The MAGNETIC FIELD pattern around a SOLENOID is the same as for a bar magnet and the polarity at the ends is determined using the letters N and S in relation to the direction of current.

Both HARD and SOFT iron can be made magnetic by being placed inside a solenoid. When removed from the solenoid, hard iron retains its magnetism, but soft iron loses its magnetism.

An ELECTROMAGNET can be made by passing an electric current through a wire which is wrapped around a soft iron core. When the current is switched off, the core loses its magnetism.

The STRENGTH of an ELECTROMAGNET depends on:

1) the strength of the electric current,
2) the number of turns of wire.

A RELAY is a switch operated by an electromagnet. It uses a small current to operate the switch in another circuit which carries a large current, eg using a computer to control a mains electric heater.

An ELECTRIC BELL (*right*) operates as follows:

1) the switch is closed and a current flows,
2) the solenoid becomes magnetic, attracting the hammer,
3) the hammer moves towards the solenoid, striking the gong,
4) the contacts open at x, breaking the circuit, stopping the current,
5) the hammer springs back, closing the contacts,
6) the current flows again and the process repeats.

A TELEPHONE operates as follows:

1) sound waves cause the mouthpiece diaphragm to vibrate,
2) the diaphragm sends compression waves through the carbon granules,
3) the waves cause the resistance of the granules to alternate,
4) the alternating resistance causes a series of current surges,
5) in the earpiece, the current surges are changed back into sound.

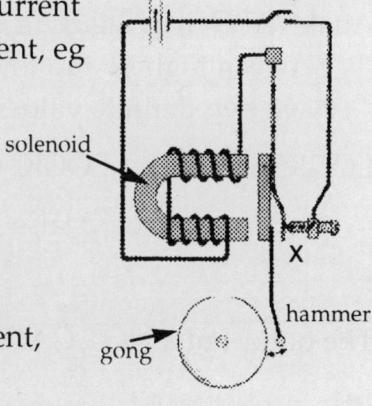

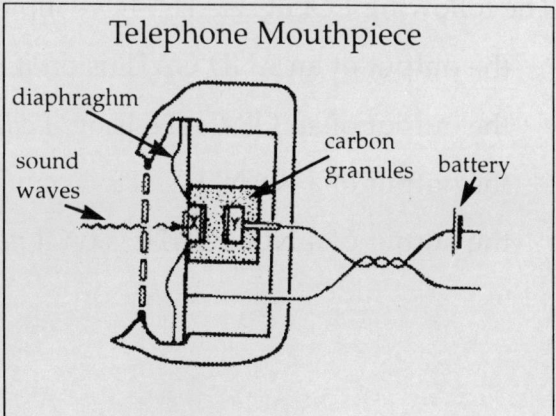

Telephone Mouthpiece

The battery sends a direct current through the carbon granules from Y to X

A current may be INDUCED in a conductor by:

1) moving the conductor across the field lines of a magnet (or vice versa)
2) changing the current in a neighbouring conductor.

A TRANSFORMER can increase or decrease potential difference. An alternating current is fed into its primary coil, and an A.C. output is induced in its secondary coil.

A MICROPHONE operates as follows:

1) sound waves cause the diaphragm to vibrate,
2) the diaphragm causes the coil to vibrate in the magnetic field,
3) a small alternating current is induced in the coil,
4) the induced current is fed to an amplifier to make it larger.

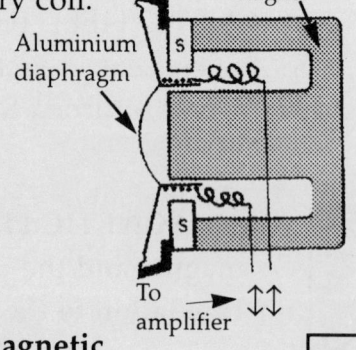

A CURRENT CARRYING STRAIGHT CONDUCTOR in a magnetic field experiences a force, and the size of this force depends on:

1) **the strength of the magnetic field,**
2) **the size of the current,**
3) **the length of the conductor.**

A SPEAKER operates as follows:

1) an alternating current is fed into the coil of wire,
2) the coil is in a magnetic field and, hence, the coil experiences an alternating force,
3) the coil makes the cone vibrate and sound waves are emitted.

An ANALOGUE SIGNAL can vary smoothly and have any value between highest and lowest, while a DIGITAL SIGNAL can only have two definite values, as follows:

LOGIC LEVEL 1 / LOGIC LEVEL 0 *or:*

on / off,   true / false,   +5 volts / 0 volts,   high/low

The output of a NOT GATE is on if the input is off.

The following LOGIC GATES have inputs A and B, and one output:

the output of an AND GATE is on if *both* input A and input B are on,

the output of an OR GATE is on if *either* input A *or* input B is on,

the output of a NAND GATE is on if input A *and* input B are not *both* on,

the output of a NOR GATE is on if *neither* input A *nor* input B is on.

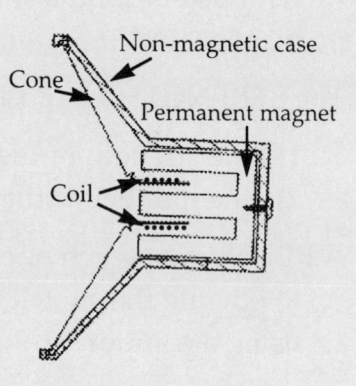

Radioactivity

(FOR DOUBLE AWARD SCIENCE, THIS TOPIC IS IN

THE THEME "SCIENCE AT WORK")

PROTONS (positively charged) and NEUTRONS (no charge) are found in the NUCLEUS or centre of an atom.

ELECTRONS (negatively charged) orbit the nucleus in shells.

In a NEUTRAL ATOM, the number of electrons equals the number of protons.

ISOTOPES are atoms of the same element with different numbers of neutrons. Each form of the element is called a NUCLIDE.

The MASS NUMBER (A) of an atom is the total number of protons and neutrons in the nucleus.

The ATOMIC NUMBER (Z) of an atom is the number of protons in the nucleus.

If N is the number of neutrons in the nucleus, then: A = Z + N.

The NOTATION for atom X is: $_{Z}^{A}X \Rightarrow {}_{Z}^{Z+N}X$

RADIOACTIVE ATOMS have nuclei which are unstable and tend to disintegrate, emitting alpha, beta or gamma radiation.

ALPHA RADIATION consists of a stream of alpha particles (or helium nuclei). Each particle consists of 2 protons and 2 neutrons, has an electric charge of +2, and a mass of (proton mass) x 4.

The symbol for an alpha particle is: $_{2}^{4}\alpha$

BETA RADIATION consists of a stream of electrons. Each particle has an electric charge of -1, and a mass of (proton mass) x 1/1800.

The symbol for a beta particle is: $_{-1}^{0}\beta$

GAMMA RADIATION is a high frequency (small wavelength) electromagnetic wave and, hence, has no electric charge and no mass.

The symbol for gamma radiation is: γ

IONISATION, a process whereby electrons break free of their atoms, is caused by radiation in materials through which it passes. Alpha radiation has a strong ionising effect, beta has a weak effect and gamma has a very weak effect. Ionisation can be harmful to living things.

Radiation can be ABSORBED by:
- ALPHA: sheet of writing paper,
- BETA: 5 mm of aluminium,
- GAMMA: 25 mm of lead reduces intensity to half (never fully absorbed).

ALPHA DECAY: when a nuclide decays by alpha emission, it becomes a nuclide with an atomic number 2 less, and a mass number 4 less than before.

$$^{A}_{Z}X \rightarrow ^{A-4}_{Z-2}Y + ^{4}_{2}\alpha$$

BETA DECAY is a two stage process whereby a neutron splits into a proton and an electron, and the electron is then emitted from the nucleus. The nuclide's atomic number increases by 1 and its mass number stays the same.

$$^{A}_{Z}X \rightarrow ^{A}_{Z+1}Y + ^{0}_{-1}\beta$$

GAMMA emission causes no change in mass or atomic number.

The HALF-LIFE of a radioactive nucleus is the time taken for half the nuclei present in any given sample to decay

ABS DOSE = $\frac{E}{M}$ = Bq Becquerel

or

The HALF-LIFE of a radioactive nucleus is the time taken for the activity of any sample to fall to half its original value. *$A \propto \frac{N}{t}$*

FISSION occurs when a neutron enters a nucleus and splits it into two smaller nuclei, releasing more neutrons and great energy.

A CHAIN REACTION occurs when the neutrons released by one fission go on to split other nuclei.

Bq Dose Sievert (Sv)

The Earth in Space

(FOR DOUBLE AWARD SCIENCE, THIS TOPIC IS IN THE
ENVIRONMENT SECTION OF THE SYLLABUS)

ACCRETION: The process whereby gravity forces between particles of matter in space draw them together to form a planet or a star.

AEROLITE: A stony meteorite.

ASTEROID: There are thousands of these mini planets orbiting the Sun, the largest of which is much smaller than the Moon.

ATMOSPHERE: The layer of gases around a planet. Earth - mainly nitrogen and oxygen. Mars and Venus - mainly carbon dioxide.

AURORA BOREALIS (or NORTHERN LIGHTS): The Earth's magnetic field directs charged particles from the Sun to the sky above the North Pole, where they cause gases in the atmosphere to emit light.

AXIS: An imaginary line through the centre of a planet around which the planet spins.

BILLION: One thousand million or 1,000,000,000 or 10^9.

BLACK HOLE: When pulsars run out of fuel, they collapse to form a black hole in which gravity is so strong that light cannot leave.

BLUE GIANTS: These are the biggest, hottest stars.

CENTRIPETAL FORCE: Gravity provides this force which keeps the planets in orbit around the sun, and moons in orbit around planets.

COMET: A dirty snowball which orbits the Sun in a path which brings it close to the Sun at times. The Sun vapourises the ice and the vapour forms a tail which is always directed away from the Sun.

CONSTELLATION: A group of bright stars which make a pattern, eg. Taurus the Bull. Twelve of the constellations are called the signs of the Zodiac.

CORE: The centre of a planet or star. The Earth's core consists of a very hot and dense metal liquid.

CORONA: The outermost layer of the Sun which looks like a faint halo and stretches millions of kilometres into Space.

COSMONAUT: The Russian word for a space traveller.

CRESCENT: The shape of the moon when less than half of it is illuminated.

CRUST: The thin, solid, rocky outer layer of the Earth

FIRST QUARTER: The phase when half of the Moon is illuminated, 7 days after a New Moon.

FULL MOON: The phase when the moon is completely illuminated.

GALAXY: A huge group of stars - up to several million. There are millions of galaxies in the Universe. We live in the Milky Way galaxy, and the nearest galaxy to us is called the Large Magellanic Cloud, 175,000 light years away.

GRAVITY: Every object in the Universe has this pulling force. The force is greater for larger masses and it decreases as the distance between masses increases.

GIBBOUS: The shape of the Moon when more than half of it is illuminated.

LAST QUARTER: The phase when half of the Moon is illuminated, 7 days after a Full Moon.

LIGHT-YEAR: Distances in space are measured in these. One light-year is the distance that light travels in one year - about 9500 billion km.

LUNAR MONTH: 28 days, the length of time it takes the Moon to orbit the Earth.

MANTLE: The layer of the Earth between the core and the crust. It consists of hot rocks which are mostly solid but some of which are liquid.

METEORS or shooting stars are streaks of light caused by tiny pieces of material falling from space and burning up in the Earth's atmosphere.

METEORITES are meteors which are too big to burn up, so they land on the Earth's surface.

MILKY WAY: The spiral shaped galaxy of 100 billion stars in which the Sun is situated. The nearest spiral shaped galaxy to the Milky Way is the Andromeda Galaxy, 2 million light years away.

NEAP TIDES: These occur when the Moon is at the First or Last Quarter phase and the gravitational pulls of the Moon and the Sun are at right angles to each other. The distance between the high tide mark and the low tide mark is at a minimum.

NEBULA: A cloud of gas and dust floating through space.

NEBULAR THEORY of the origin of the Solar System: This states that the Sun was formed from a nebula by accretion. The Sun then threw out material from which the planets formed.

NEUTRON STAR: The very dense surviving part of a blue giant star after a supernova explosion.

NEW MOON: The phase when the Moon is invisible, as it is between the Sun and the Earth.

ORBIT: The curved path of an object that travels around a star or planet. Orbits tend to be elliptical, rather than circular, in shape.

P-WAVES: Longitudinal shock or seismic waves, caused by an earthquake, which travel through the Earth.

PLANETS: Bodies which orbit a star. In our solar system, the four inner planets (Mercury, Venus, Earth and Mars) are small and rocky, while the next four (Jupiter, Saturn, Uranus and Neptune) are large and gaseous. The furthest planet, Pluto, is small and rocky.

PLUTO: This planet has an eccentric orbit which brings it closer to the Sun than Neptune at times.

PULSARS: Neutron stars which emit radio signals.

RED GIANT: When a star about the size of the Sun runs out of fuel, it swells 100 times and emits red light for several million years.

RED SHIFT: Light from distant galaxies is redder than it should be because they are moving away from us (ie the Universe is expanding).

S-WAVES: Transverse shock or seismic waves, caused by an earthquake, which travel through the Earth.

SATELLITE: Anything that orbits a planet. Moons are natural satellites.

SIDERITE: A metallic meteorite.

SOLAR SYSTEM: The Sun, the nine planets with their, roughly, 50 moons and the thousands of smaller bodies which orbit the Sun make up the Solar System.

SOLAR WIND: Charged particles which stream away from the Sun in all directions.

SOLAR YEAR: 365¼ days - the length of time which it takes the Earth to orbit the Sun.

SPRING TIDES: These occur when there is a New Moon or a Full Moon and the gravitational pulls of the Moon and the Sun are in line. The distance between the high tide mark and the low tide mark is at a maximum.

SUPERNOVA: The explosion which occurs at the end of the life of a blue giant.

STAR: A ball of gases which produce energy by the process of nuclear fusion (it is not true to say that the gases burn). Stars can evolve as:

1) blue giant → supernova → neutron star → pulsar → black hole, or

2) star (like our Sun) → red giant → white dwarf → black dwarf.

SUN: The star at the centre of our Solar System. It is about half way through its life of 9 billion years. One million Earths could fit inside the Sun. It takes light 8.5 minutes to travel 150 million km to the Earth from the Sun.

SUN SPOTS: Dark spots on the Sun which may last from several hours to 200 days. When the Sun has very few sunspots, our climate tends to get colder — the Maunder Minimum.

UNIVERSE: The whole of Space and everything in it.

WANING: This occurs between the Full Moon and New Moon phases, when the illuminated part of the moon becomes less from night to night.

WAXING: This occurs between the New Moon and Full Moon phases, when the illuminated part of the moon becomes greater from night to night.

WHITE DWARF: After several million years, a red giant shrinks to become a white dwarf. It turns yellow, orange and then red and finally fades away to become a black dwarf.